Charles B. (Charles Boyd) Kelsey

The office treatment of hemorrhoids, fistula, etc. without operation

together with remarks on the relation of diseases of the rectum to other

diseases in both sexes, but especially in women,

Charles B. (Charles Boyd) Kelsey

The office treatment of hemorrhoids, fistula, etc. without operation
 together with remarks on the relation of diseases of the rectum to other diseases in both sexes, but especially in women,

ISBN/EAN: 9783744738743

Printed in Europe, USA, Canada, Australia, Japan

Cover: Foto ©berggeist007 / pixelio.de

More available books at **www.hansebooks.com**

THE OFFICE TREATMENT

OF

HEMORRHOIDS FISTULA, ETC.

WITHOUT OPERATION

TOGETHER WITH REMARKS ON THE RELATION OF
DISEASES OF THE RECTUM TO OTHER DISEASES
IN BOTH SEXES, BUT ESPECIALLY IN WOMEN, AND
THE ABUSE OF THE OPERATION OF COLOSTOMY

BY

CHARLES B. KELSEY, A.M., M.D.

LATE PROFESSOR OF SURGERY AT THE NEW YORK POST-GRADUATE MEDICAL
SCHOOL AND HOSPITAL; FELLOW OF THE NEW YORK ACADEMY OF
MEDICINE, THE NEW YORK COUNTY MEDICAL SOCIETY, ETC.

E. R. PELTON

No. 19 EAST SIXTEENTH STREET, NEW YORK

1898

PREFACE

Of the three lectures contained in the following pages one only has previously appeared in print, and that the last.

In presenting them in this form the author has done what is in his power to counteract the far too prevalent idea in the medical profession, that the only radical treatment of these affections is by operation; as well as to call attention to what he considers the far too frequent misuse of one of the most important of all the operations.

C. B. K.

No. 18 East Twenty-ninth Street,
New York, 1898.

CONTENTS

THE CURE OF HEMORRHOIDS, FISTULA, FISSURE, AND OTHER AFFECTIONS OF THE RECTUM BY OFFICE TREATMENT, WITHOUT OPERATION.

Gentlemen: You have just seen me perform what I consider the best of all the well-recognized surgical operations for the cure of hemorrhoids. As a consequence some of you, if you follow the usual rule, will at once purchase the instruments you have seen used, and when you get back to your patients you will adopt this as your routine practice.

To the surgeon nothing is as satisfactory as a successful surgical operation. It is positive and definite in its results; the patient is cured in the shortest and (for the surgeon) the easiest way possible; and the time and

9

expense necessary, as well as the amount of suffering and danger involved, can often all be accurately estimated before its performance.

To the patient, on the other hand, nothing is much less satisfactory than this plan of treatment. To her it is attended by many horrors, exaggerated of course, but none the less real. You all know this, and you are in the habit of arguing it down with your timid patients, and pooh-poohing it, and trying to convince them that the taking of ether is rather pleasant than otherwise, and dilating on the benefits to be received; and yet I know of no class of men who will suffer longer to avoid it than our own profession. How many of you would be glad in your hearts were you at this moment where this woman is, cured of your hemorrhoids and on the road to a safe and speedy recovery, and yet how many of you have the courage she has had to submit to on operation to accomplish this result? None of you will do so until you are driven to it by suffering and annoyance.

Several years ago I had what was an annoying as well as an amusing experience with a lady patient growing out of this very feeling. She was suffering severe pain with her hemorrhoids, and came to me for relief. I advised operation, and, although she was much upset by the idea and very nervous, she consented. Only a few hours elapsed before she was again in my office, earnestly begging for a cure without an operation, and I went over the whole question of the advantages of the clamp and cautery and the dangers of carbolic acid a second time. She was convinced for the moment and the operation was to be on the next day, but her heart failed her and again she came, this time with a friend, and insisted that I should cure her without operation. She was convinced that I must be able to do it; that for some reason I was unwilling to do it: and as I persisted that I knew of but one method except operation (that with injections), as was indeed true at that time, and that I considered that plan so unsafe I was unwilling to employ it, we parted. Only a

short time after I received a letter full of bitter
reproaches, because I had been brutal enough
to cause her such mental agony by even sug-
gesting a surgical operation; and saying she
had been entirely cured by a certain irregular
practitioner who had used injections. She
was not cured, but she was greatly relieved for
a time, and, at all events, she was happy. Of
course she was absolutely illogical, and took a
great risk in her ignorance of surgery and
might have come to grief; but I have always
had a certain amount of sympathy for her side
of the question, and perhaps that one case
more than any other has kept my mind for
years intent on the question of curing these
cases without forcing them to submit to what
they so much dread.

I was much struck some time ago by the
statement of a friend who has devoted his life
to diseases of women, that there was no treatment for these affections except operative
treatment, and that his own office was not a
place for treatment, but merely for examination and for making appointments for opera-

tions. Such a practice is indeed most satisfactory in its results, and the results are reached in the shortest and surest manner. There is in fact but one objection to it, and it is not at all strange that that objection should come from the patient.

Habits of thought and of medical practice grow upon practitioners. Moreover, most men are merely imitative in their practice. They see, they read, and follow suit. The results of operative surgery during the past few years have been wonderfully brilliant, and the facilities for teaching the art have been multiplied an hundred fold. An overcrowded profession has also been very quick to take advantage of these facilities, and the result has been a wave of operative surgery, spreading all over the land. Whether the study of the science of surgery has kept pace with the practice of the art, or whether, in diseases of women, for instance, a man may not occasionally be found who, though expert enough to remove the ovaries, has never learned to diagnosticate their condition before operation, is a matter we will not discuss. This is a surgical clinic, and you gentlemen are here to learn how best to operate upon diseases of the rectum.

But, is it not possible that in the desire to learn how best to operate, and in seeing me operate as often as you do, the thought of how much might be done for these patients without bringing them into the hospital and placing them on the operating-table, and the methods of cure without these accessories, may be lost sight of? Have you noticed that in these less serious cases of rectal trouble upon which you see me operate, the patients are almost invariably either working-people, or, at least, people who can only devote a few days to getting well? Such is the fact. Many live at a distance and must get home as soon as possible, and many others are poor people who cannot devote two or three afternoons each week to waiting their turn in the dispensary for medical treatment, even to avoid an operation. For this reason many cases which would be curable by patient office treat-

ment are at once put into the wards and
brought before you for operation simply to
save time.

In this branch of surgery this question between operative and non-operative methods
of treatment has always been kept alive by the
existence of two distinct classes of practitioners—surgeons who *could* operate and pre-
ferred to do so for the good of their patients,
and irregular practitioners who vaunted secret
and non-surgical methods of treatment without operation. It is also possible that the former have become so accustomed to looking
only to operative methods that they have acquired the mental habit of asking themselves
in the presence of every new case, What can
be done for this patient by operation? and
have not cultivated the non-operative side of
the specialty.

That there is a non-operative treatment for
these diseases is not to be disputed. I do not
at all mean that the quacks are in possession
of it and we are not, or that we can learn anything from them in surgery, for this I do not

admit or believe. I simply mean that a very large proportion of all cases of piles, fissures, superficial ulcers, and pruritus; and a certain proportion of abscesses and fistulæ may be radically cured in one's office without resorting to ether, or confinement to bed. The words *radically cured* are used advisedly, because to say that any man or woman can be temporarily benefited in this way would be only to repeat what all the world knows already.

The methods by which these results may be accomplished do not include any new surgical principle, but simply the application of modern methods more effectively and perhaps with greater ingenuity. These methods have, it may be said, been floating through surgical literature for years, but have received only temporary and casual attention. No man seems to have deliberately tried them to their full efficiency as a rule of practice, before resorting to the better known and generally accepted operations under anæsthesia.

Take for example the question of fistula.

We all know that many of them cannot be made to heal without laying them open into the bowel, and that the easiest and best way to do this is to put the patient under ether and use the knife. From this fact has grown the general practice of operating upon all fistulæ without spending any time over less certain plans of treatment.

And yet many years ago Allingham wrote that he had cured many cases of fistula by inserting a perforated collar button into the cutaneous orifice for dilatation and drainage. The button was merely his own ingenious device. The essentials were the dilatation and drainage.

Take now the following case: A gentleman consulted me last October for a fistula, resulting from an ischio-rectal abscess, in other words, one that ran far up into the fossa, and had its cutaneous orifice two inches from the anus. He had been operated upon four times without cure, and already had considerable incontinence of fæces as a result. Another operation he stated plainly he would not submit to for fear of still greater damage to the muscle, and if I told him he could not be relieved without operation he had deliberately determined to suffer the ills he had. The sinus had already been curetted several times without benefit. Promising only to do what I could, I first thoroughly dilated the old sinus its entire length, and then simply introduced a gauze drain. In three weeks, a part of which he spent on the railroad, attending to business, he was entirely cured, and he has remained so; and yet had he not dictated his own treatment I should probably have operated in the usual way.

There is certainly no new principle involved in treating old sinuses connected with the bowel by dilatation, stimulation, cauterization, and drainage, and many may be induced to heal by these measures. Usually, I think, they will be the straight ones without branching offshoots, and where no tubercular taint exists. But although there may be nothing new in the principle, its successful and painless application to this class of cases in office practice will always remain a matter of personal ingenuity, experience, and delicacy of touch.

What was accomplished here was exceptional, although there have been men in the past (irregular practitioners) who devoted themselves entirely to the cure of these cases in this way, and, I am informed, with considerable success. But what was perhaps exceptional in an old and extensive fistula need not be so in less severe cases or in pruritus, fissures, superficial ulcerations, and hemorrhoids; and although the giving of ether and confining the patient to bed for a surgical operation may be the shortest and perhaps the most satisfactory way of treating these cases, there are none of them that cannot be cured by office treatment much more satisfactory to the patient.

The treatment of piles without operation has been the battle-ground between the surgeons and the irregular practitioners for years. The whole question of the injection of irritating fluids, the best known of which is carbolic acid, has been thoroughly discussed. The

weight of experience conclusively shows that
serious accidents are always liable to follow
the depositing of any irritating fluid with a hy-
podermic syringe into the middle of a vascular
tumor like a hemorrhoid, with the idea that
the resulting disturbance will be just sufficient
to cause consolidation and not enough to
cause suppuration. If this treatment were free
from the risk of abscess and resulting fistula
it would leave little to be desired, but it never
can be, and it will therefore remain an uncertain and dangerous method, bound to bring
dissatisfaction to both patient and surgeon in
a considerable proportion of cases, and not
radical in its results.

And yet, knowing this, as both the profession and the laity do, the treatment still re-
mains popular because of this deeply rooted
fear of ether and the knife, and when it turns
out badly the patient is generally worse off
than before.   I refer to it merely to exemplify
the popular demand for non-operative treatment, but, after a good deal of experience with
it, cannot advise it.

I have taught for years that a straightforward surgical operation under ether was better than this method, because, although it was disagreeable, it was free from risks, and, therefore, you never see me practise it. And this rule applies to all forms of disease; for in trying to avoid the well-known operative methods with their accessories which the patient dreads, the practitioner must not allow himself to be drawn into uncertain experiments and plans of treatment liable to be followed by disaster. This would certainly not be a kindness to the sufferer, however charitably intended.

You all know the time-honored treatment for painful ulcers or fissures at the anus. It is to forcibly stretch the muscle by introducing the fingers or thumbs of the two hands and then separating them as far as possible. To do this, unless with an absolutely brutal disregard of pain, requires ether or chloroform. It is very effective and seldom fails to cure, but have any of you ever seen me resort to it? You have not, and for the reason, I suppose, that I have never been able to overcome in my own mind the sense of the ludicrous which the disproportion between the disease and this method of cure has always excited. Somehow I always think of it as one does of killing a fly with a ·club. It is certain and speedy, while another plan of treatment may take two or three weeks; but offer the patient her choice between them and see which she will prefer.

For what are these fissures after all? Often only abrasions, and only important on account of the severe pain due to their location. Even when they are older and deeper, with hard, raised edges, the same treatment which will cure an ulcer elsewhere will cure it here.

Some of you may also have noticed the rarity with which cases of slight ulceration are brought before you for operation, and this is because until they become very far advanced they are always treated by non-operative measures and local applications. Indeed the operations when necessary are so severe that patients can only be induced to submit to them

when every other method of treatment has
been exhausted, and the average man does not
care to undertake them. To treat them properly before operation requires, however, every
facility for illumination of the bowel, and the
proper instruments for exposing them, combined with a good deal of skill and experience
in the choice of local applications.

In every case in which you have seen me
operate for destructive ulceration of the bowel
it is safe for you to presume that years have
been spent in previous non-operative treatment by many men, which has been unsuccess-
ful; and when the operation comes it is perhaps the most difficult, certainly one of the
most difficult, in surgery.

With regard to some of the other more common affections, pruritus or itching for example, you know how seldom it is necessary to
etherize the patient and put her or him
through a surgical operation in order to effect
a cure. The only indication for operation at
any time is to overcome the thickening of the
skin which is sometimes present, and this can

be done in time without resort to etherization and the sharp curette.

Thus it happens that you will find yourself confronted in nearly all these cases with two different methods of cure. One is by an operation, the other by office treatment. The former may be the shorter and for you the easier and more satisfactory, the latter will be much more acceptable to the patient, provided he is in condition to devote to it the necessary time. The results will be the same in either case, and my object has been for the moment to draw your attention away from the operations in which you are all so much interested, and to keep you from being carried away with the idea that operation is always necessary. For in this matter the patient's sensibilities have a right to consideration and your own preferences should not be allowed too absolute sway.

One of my assistants here who has helped me in many operations, and who really has acquired considerable facility in the merely operative branch of the work, told me, after spending a day or two in my office, that he had never had the least idea of what the daily routine of office examinations and treatment consisted in. To him up to that time the whole subject had been included in the two words, first diagnosis, second, operation. Cure without operation was to him practically unknown, and I sometimes fear that it may be so with you, for the reason that this is a kind of work impossible to show you here. In fact, although you see an abundance of operative surgery and a good deal of it of a major kind in connection with these cases, you do not see much that constitutes the daily office-work of a practitioner, or have a chance to acquire the facility in examination, diagnosis, and office-treatment, which are essential to cure.

One reason for this is that great gentleness is absolutely necessary in these examinations to avoid inflicting unnecessary pain, and, perhaps, doing worse, and inflicting serious harm; and I have never considered it fair that patients who come here, suffering intense pain as they sometimes do, should be subjected to repeated examinations, which, when made in the gentlest and most skilful manner, may be necessarily very painful, merely for the sake of practice for the student.

Again, a diseased bowel is often a very delicate thing to handle, and fatal accidents have occurred from this cause combined with rough manipulation.

But just as the oculist has his ophthalmoscope for looking at the retina, and the throat specialist has his light and mirrors, and the surgeon has his cystoscope for examining the interior of the bladder, so have instruments for looking into the bowel always been used, and recently they have been improved and rendered more efficient, so that instead of being limited in our treatment to the lower three or four inches of the bowel, as practically we have been in years gone by, we now see twelve or sixteen inches with facility.

This has greatly increased our scope of accurate diagnosis and brings within the field just that part of the bowel which was before the most difficult to examine by any method. For the bowel as it lies in the abdomen, before dipping down into the pelvis, may be reached and examined with the fingers pressing upon the abdomen, and as it leaves the pelvis and comes to the skin it has always been examined either with the finger or by such instruments as we have possessed; but the part which lies between these two, and which constitutes a loop several inches long, has always been a *terra incognita* to the examiner, because it could not be reached either from above or below. This has been in a great measure overcome by the use of modern instruments, with the handling of which it is necessary that you should render yourself perfectly familiar.

I consider this advance one of the most important that has been made in late years, for by it a whole group of cases which formerly could only be guessed at, and were even then practically out of the reach of local treatment, has been brought within the range of vision and direct treatment, and these cases are some of the most serious which present themselves.

Let me give you an example which made a strong impression upon my own mind years ago, and has always since had an effect upon my practice in at least inculcating caution about expressing an opinion as to this class of cases.

The patient was a young member of our own profession, a man of wonderful physique and apparent perfect health, who came to me from a long distance for an opinion as to two slight symptoms referable to the bowel. One was occasional slight traces of blood in the passages, which were otherwise natural in form and frequency; and the other was a sense of pain and uneasiness in the left side of the abdomen low down, which he referred to this part of the intestine. He was worried about his condition, although it seemed scarcely with reason, and the most careful examination by the methods then known, including conjoined manipulation under ether and the use of a full-sized exploring bougie, revealed nothing but some slight hemorrhoids. To clear up the diagnosis these were removed, not because they caused any annoyance, but simply to eliminate one possible source of the blood in the passages. There was no change in the symptoms, and the diagnosis became impossible without waiting developments or opening the abdomen for exploration. As this seemed both to him and me hardly justified by the symptoms, which might easily be due to a simple cause, he decided to return to his home and wait the gradual unfolding of the case; and this decision cost him his life, for he was suddenly seized, a few weeks later, with complete intestinal obstruction and died before surgical relief could reach him. On autopsy a small ring of cancer was found just in this part of the bowel, completely closing it.

This is a sample of the cases the nature of which we could formerly only surmise in their early stages, but now brought within the range of exact diagnosis by new methods of examination.

But not only are we helped in diagnosis very materially, but, what is much more gratifying to the patient, we are enabled by direct applications, through these instruments, to cure a large number of cases otherwise incurable. I have removed in my office and without ether, polypoid growths through these instruments at eight and ten inches from the anus with the same accuracy as could be done in the larynx, and have cured by local applications extensive, though superficial, ulcerations incurable by any other method.

When you think that many patients are now going about who have been subjected to the frightful deformity of an artificial opening into the bowel in the left groin for just such conditions as these, you will appreciate the value of this method of treatment.

You see here a set of these instruments of different sizes, and in suitable cases you will have an opportunity to see them used. But to accomplish anything with them, except to cause the patient unnecessary pain, you must not only have every facility, but considerable practice. These are modifications of the original ones, devised by myself for the purpose of obtaining a wider view and a larger field for manipulation. To use them at all is at first difficult, to see anything clearly with them is much more difficult, and to be able correctly to interpret what can be seen requires special skill.

The matter of light is of great importance, for these examinations are best made by artificial light in a dark room, and you will find electric light possesses many advantages over any other. Many cases can be seen perfectly well by ordinary daylight, but when you come to this more delicate work high up in the bowel, you need a very concentrated light thrown on the exact spot, which can only be secured by the use of a forehead mirror, and by an arrangement for changing the focus instantly over a range of many inches.

This practical use of electricity has until recently been nothing but a constant annoyance to surgeons, because of the difficulties of its application. Batteries of all kinds have been tried again and again, and the universal experience has been that they never worked when needed. But now that nearly every house can be supplied from the street it can be adapted to office-work, both of examination and treatment in the most satisfactory way. All this implies expense on your part for apparatus and modern instruments, but the results obtained will more than repay you. But it implies something much more important than expense, and that is skill, practice, and great delicacy of touch, and these are things which cannot be taught, but which each one must acquire for himself.

If now, gentlemen, I have said enough to counteract the general impression which you all have, and which taking a course here in the operative surgery of the rectum tends to foster, that there is no successful and radical treatment for these diseases except by ether, operation, and confinement to bed, I shall have accomplished my object.   Beyond this point you must decide which plan of treatment you will pursue in any individual case for yourselves.   As far as possible I shall endeavor to teach you both, but I have learned by experience that most men take far more interest in the apparently more brilliant operations than in the patient and less exciting, though equally successful non-operative treatment. With myself the case is the opposite, for, having operated all my life, I have come to enjoy the demands which office-treatment makes upon one's ingenuity and manipulative skill, and the successes which sometimes follow even in what appear at first sight the least promising cases. Many of the more serious affections cannot be cured in this way, and the treatment is one better adapted to those in comfortable circumstances than to the poor; but much more can be done than you imagine. This is not the time to attempt to describe the various plans of treatment of all the different affections which are curable by these methods, in fact, that would demand a special book of itself. Much of the manipulation cannot be learned from description, any more than you can learn how to use an ophthalmoscope or laryngeal mirror effectively, or to remove a growth from the vocal cords, or one of the turbinated bones from the nose. It is emphatically a kind of practice to be studied while in personal contact with the patient.

ON THE RELATION BETWEEN DISEASES OF THE RECTUM AND OTHER DISEASES IN BOTH SEXES, BUT ESPECIALLY IN WOMEN.

Gentlemen: In women the greatest obstacle to the success of any form of treatment of diseases of the lower bowel, either operative or non-operative, will often be found to be the co-existence of some form of uterine, ovarian, or bladder trouble.

Unless this is borne in mind before commencing treatment, you will be more than likely some day to have it disagreeably impressed upon you. For some lady will consult you for a cure for her hemorrhoids, and she will, in all probability, be a patient with whom you are most anxious to acquit your-

self creditably; and you will operate and as-
sure her she will be all well in about a fortnight, but the wounds you make will not heal.
Then, indeed, your trouble will begin, and
after weeks of the use of salves and lotions,
suppositories, and applications in your office,
you discover, or some other doctor discovers
for you, that the woman has uterine disease
of some sort, and that the wounds are not
likely to heal until that also is relieved.

Where the uterus is enlarged or misplaced,
or the patient is a chronic sufferer from ovarian pain, or irritation of the bladder, any
wound of the rectum may refuse to heal kindly
and result in a sluggish, unhealthy sore, and
treatment may give little relief, the cause of
the trouble remaining constant and counter-
balancing any attempts at cure.   It is for this
reason that I have so often tried to impress
upon you that intelligent and successful treatment of affections of the lower bowel was impossible by any man who did not also understand the other diseases of women and was
not prepared not only to make a diagnosis of

their presence and nature, but also intelligently to treat them. In women the two branches of practice are inseparable.

The same statement applies to many affections of the urinary organs in men, such as enlarged prostate, and irritation of the neck of the bladder.

Irritation of the bladder with frequent urination being one of the well-recognized causes of piles, how can one hope to cure the latter by any kind of treatment while the former is neglected? In all these cases the co-relation of symptoms, due simply to anatomical causes, is so marked that unless the surgeon is versed in the treatment of both he had better call in professional help.

I have here a case which illustrates this point in a remarkably forcible way, and I will use it to still further impress it upon you.

This woman came to us some months ago complaining of only one symptom, which was pain in the lower bowel after having a passage. She was otherwise in perfect health as far as she knew, but this one thing had made her a semi-invalid for years, and having exhausted every means of relief she was pretty nearly hopeless. How thorough her efforts had been may be judged from the fact that she had been operated upon for this trouble with the rectum by three of the best general surgeons in New York, and had been examined by one specialist in diseases of women, who had failed to detect any trouble with the internal organs.

An ulcer, or a hemorrhoid, or something, I have forgotten just what, had been removed by one, and this giving no relief to the pain, the scar left by that operation was cut out by a second, and the third, at his wit's end to discover the cause of the trouble, had etherized her and done something else, and as a result of all these efforts in her behalf she had not only the same old pain as bad as ever, but lack of control over the passages from frequent cutting of the muscle which closes the anus. Don't imagine this was due to any carelessness, because the patient was not then poor, but abundantly able to command the best, and she had the best.

The most careful examination of the rectum showed it to be free from anything which could cause the pain, and a little further examination showed a very much enlarged uterus and a prolapsed right ovary. The act of having a movement of the bowel caused pressure upon the misplaced ovary resting upon it and caused her the pain which kept her in bed several hours and then gradually wore away.

When this condition was explained to her and her husband it was received with absolute incredulity. She could not understand, as you can, how a pain in the rectum should come from an entirely different organ; but she was cured by reducing the size of the uterus and replacing the ovary by the proper treatment; and she tells me now that for the last few months she has been comparatively well for the first time in years.

I bring this case before you not as a curiosity, but as a type of a common class you cannot possibly fail to meet, although seldom with such a marked history. You will constantly be consulted by women who complain of trouble of some kind in the rectum, and in whom the cause of the pain, or the hemorrhoids, or the fissure, whichever it may be, is in the uterus or ovaries.

One of the first steps, therefore, which must be made in fitting one's self for the practice of this specialty is to learn to diagnosticate the condition of the female pelvic organs by conjoined manipulation. You know what an amount of study this implies, and what years of practice and special training are necessary before it can be accomplished. None the less it must be acquired if you would take rank to-day as a surgeon, and not be content to be merely a pile-doctor, and not always successful even at that.

You will find, however, that after once acquiring facility in examination of the lower bowel with the index finger you have made a decided step toward intelligent examination of the other organs contained in the female pelvis. For were the examiner limited to one method of doing this there is no question but that conjoined examination per rectum will yield the most accurate results in a large majority of cases. This is well recognized by our gynæcological friends, and mention of it is never omitted from their text-books, most of which also contain chapters on the treatment of diseases of the rectum itself, showing the necessary relation of the two branches of surgery in women.

Not only is it absolutely essential that the practitioner who desires to be successful as a specialist in diseases of the rectum should know how to diagnosticate the condition of the other pelvic organs, but there are many diseases of those organs which he must know how to treat and cure unless he is constantly willing to confess to those who consult him that his powers as a surgeon are of very limited scope.

A lady consults you for hemorrhoids, for example, and you find they are either caused by or greatly aggravated by the presence of an enlarged or misplaced uterus. If you are content to say, " Madam, if you had hemorrhoids and no disease of the uterus I might be able to cure you, but this case is entirely too complicated for me, and you had better consult somebody else," well and good. That, then, is your chosen field of work and you have a perfect right to confine yourself to it. But you are not a rectal specialist, you are merely a specialist who treats *uncomplicated* cases of rectal trouble and sends the others to somebody else, and a good many men would not be satisfied with such a position.

Should you, however, do the other thing and operate upon the hemorrhoids without paying any attention to the condition of the uterus you do a great injustice to your patient, because, although she may have come to you for the hemorrhoids, she has in reality come to you for the purpose of being made well, and no treatment you may give the hemorrhoids alone will make her well.

How well do I remember the first lady I ever operated upon for a laceration extending

completely through into the rectum, the result of child-birth! The case occurred many
years ago and had an effect in shaping my
whole surgical practice ever since, for, although I did operate, and cured her, I had no
idea when the case came to me that it was a
case of laceration, and had I known it I should
not have had the courage to undertake it; for
it was by no means a charity case, or in dispensary practice, but a case where failure would
have done me great harm, and I had never
even seen the operation performed. The case
was this, and from it you may learn my own
idea of what you must fit yourselves to do:

A gentleman walked into my office one day
with the remark that he had heard I had cured
a patient who suffered from loss of power to
control the passages from the bowels, and
asked if I thought I could cure another. Being told that I did he stated the case as fol-
lows: The lady in question had been lacerated
in child-birth, but after three operations had
finally *been cured*. Subsequently, however,
she had developed a chronic diarrhœa, and

with this had come on a falling and protrusion of the bowel, and for years when the diarrhœa was present there was no control over the passages. The case seemed plain enough and I at once operated for the prolapse, supposing that the diarrhœa and loss of power were merely the result of this condition, as is often seen.

The prolapse was cured, but the diarrhœa and loss of control remained the same, and the patient was little better than before.

I was much chagrined, and made a more careful examination only to discover that the old laceration had never been cured. Although the skin had been drawn together and had united perfectly, the muscles of the anus had been entirely left out of the operation. In other words the prolapse and lack of power were both due to the old laceration, and the loss of control was not at all due to the prolapse as I had supposed. According to all the practice of that time the case was therefore one for the specialist in diseases of women and not in diseases of the rectum, none the less it had come to me solely for the trouble with the rectum. The relation between cause and effect was of no interest to her. What she wanted was to regain control over the bowel. I therefore operated again, as I say, for the first time in my life, and with a very exaggerated idea of the difficulties, and cured the patient; and I have been operating in the same class of cases ever since, and if you intend to cure a considerable proportion of the cases that will come to you complaining of trouble with the lower bowel, you must fit yourselves to do the same.

There are many other points in which these two specialties so overlap each other as to include the same kind of cases, for most of the diseases of women cause symptoms connected with the rectum. Among these are tumors of the uterus or ovaries pressing upon the bowel; pelvic abscesses opening into and discharging through the bowel; false passages existing between the bowel and the bladder or vagina, etc. A large class of cases is those of lacerations, which, though not extensive enough to

tear completely into the rectum, still weaken
it so as to cause the most obstinate constipation from lack of muscular power to expel its
contents. Many such will come to you for
help, and it is your business to be able to cure
them just as skilfully as though the difficulty
in having a passage were due to a closure of
the bowel instead of a weakening and dilatation.

The patient's feelings must be consulted
here. You are supposed to be able to cure
all troubles of the bowel which are curable.
For that reason she comes to you, submits to
an examination, pays you your fee, and ex-
pects to be helped. When you say, "To be
sure this is a disease of the rectum, but it is not
the particular kind of disease I am in the habit
of treating; you had better go elsewhere," you
make yourself ridiculous, and she has a right
to feel that in a certain way she has been im-
posed upon, for the distinction you make so
carefully is apt to be too fine for her appreciation.

What has been said about the relation of

these diseases to those of women applies also in a measure to that of other specialists and surgeons, but not to so great an extent. A large class of cases will be found in the bladder troubles in men, many of which will first come to you for symptoms connected with the rectum; for pain in this part is one of the most constant signs of these troubles. You may either try and treat them or refer them to somebody with special training as you feel yourself competent; but, in any event, you must be able to recognize them and know what kind of treatment is needed and not put yourself in the position of treating what may be only a single symptom of a complicated chain of disease.

As to the major operations in abdominal surgery you will constantly be called to face them, and you must be competent to perform them or at once hand them over to one who is. Would any of you like to admit to a patient that although your specialty called upon you to be able to operate upon an abscess opening into the bowel at one point, you did not feel competent to operate upon another opening a little higher? And yet, to do this, you must be ready to encounter the most difficult and dangerous cases met with in abdominal surgery. Are you equal to the resection of a stricture at six inches up the bowel and not to one at twelve, and can you treat a simple prolapse but not a prolapse due to invagination of the intestine?

Do you wish to take the position that you are a specialist in making an opening into the bowel, but cannot close one when it has been made; that you can suture the bowel to the skin, but cannot suture one cut end to the other; that it is your business to remove tumors from the rectum which are pressing upon the uterus, but not tumors of the uterus pressing upon the rectum; that it is your special business to relieve intestinal obstruction due to strictures within reach of your finger, but not from strictures an inch higher up; that your specialty includes all operations for opening the abdomen for telescoping of the bowel, but none for the same thing due to twisting of the bowel on itself? These distinctions may be satisfactory to you, but they are at least very artificial.

The public believes in specialties and demands specialists, but such fine-drawn distinctions as these are beyond their appreciation, and tend to make the whole question appear ridiculous.

So you will find the path of practice opens up into many broad fields of study and interest, and my advice to you is to follow them for your own sake. For there is danger in this path of strict specialism of dropping behind in the general advance which is going on around you; of losing your interest in the broader scientific work of the day and becoming merely a machine, practising what you learned years before and going little further. If your field of work is limited strictly to the diseases of six or eight inches of the bowel itself so will your thoughts be and your reading; but if you are determined that anything at all relating to the bowel or in any way affecting it is a part of your chosen work, then you have a specialty broader perhaps than any other, and certainly affording ample scope for all your powers of study and thought, and all your skill as surgeon and physician.

This has always been my own conception of what properly belonged to this branch of surgery, and for the sake of your own intellectual activity I should advise those of you who mean to follow it to consider the subject very thoroughly.

If you agree with me there is hope that some time I may for the last time have to answer the oft repeated and always amusing question, How long does it take and what instruments are necessary to be a specialist in these diseases? For you will then appreciate that a six weeks' course of lectures is rather too short, and your lifetime will not be too long for study; and that you will never reach a point from which you may not advance to a higher.

It is a fact that many men who have at one time devoted themselves very closely to diseases of the lower bowel have later extended the field of practice to other work which nat-

urally came from it. Allingham, senior and
junior, are both surgeons to general hospitals,
so was Van Buren. Emmet began as a rectal
specialist, and because he could not practice
the one without the other became a gynæcol-
ogist. This it is not necessary for you to do.
Take the title of a rectal specialist, if that is
your field of work; it is fully as good as gynæ-
cologist, and if you include in your work the
cases of diseases of women and the general surgery which you will be constantly brought into
contact with, the field will be just as broad.

Somebody has defined a well-educated man
as one who knows all about one thing and
something about everything. Although you
may not in medicine be able to reach the latter half of this definition, you certainly need
not limit the first part to about half a dozen
affections of six or eight inches of intestine.

ON THE ABUSE OF THE OPERATION OF COLOSTOMY, OR THE FORMATION OF AN ARTIFICIAL ANUS.

Gentlemen: We have here a typical case of most serious disease, which will serve admirably for a text for something I have been waiting to impress upon you for some time.

You see a young woman (and these cases are generally in women) who still seems in fairly good physical condition though she has been a chronic sufferer for years, and has reached a point where she feels she can no longer endure what she has borne so long. She has chronic inflammation of the lower bowel, which has gone on through all its stages of ulceration, thickening, and contraction until now there is scarcely any passage left, not enough in fact for me to introduce the tip of my little finger.

The symptoms need not now detain us. You are familiar enough with such cases to know that for years she has been passing blood and slime with constantly increasing frequency; that the bowel has been gradually closing so that there has been constantly increasing difficulty in having a passage; and we find her now in a condition of chronic intestinal obstruction with the abdomen distended and tender on pressure. What this has meant to her in the way of suffering and loss of sleep and inability to walk or work, and loss of control over the passages you can also imagine from the fact that after ten years of trying first one doctor and then another; and after having been operated upon several times by cutting and stretching the contraction which is closing the bowel, she finally throws herself completely on our hands, and, at any risk, asks for relief. The history is only too

familiar to you, for these cases are not un-
common.

So much for her side of the case. Now,
what shall we do for her? There are two
totally different ways of treating her; by one
of which we relieve, and by the other cure her;
and, when I say that the answer to this question as to the proper treatment of these cases
is to my own mind the most pressing one connected with this whole line of surgery at the
present moment, you will be able to judge of
its importance.

This patient tells us she has already been
operated upon several times. These operations have consisted in efforts to overcome the
contraction of the bowel by either cutting
through it or stretching it, and such efforts
never give more than a very temporary relief,
for the stricture immediately returns.

Recognizing the futility of all these measures Bryant, about fifteen years ago, advocated something much more radical, which
meant their practical abandonment. His
teaching was that as long as the condition was

incurable, and the bowel could not be restored to a healthy state, it might as well be abandoned, and an artificial opening made through the abdomen for the escape of its contents, at a point above the disease. By this means the patient would at least be relieved of the pain and the difficulty of having an evacuation, and would quickly regain strength, although the disease remained the same.

The teaching was bold, it meant a distinct surgical advance over what had gone before, and was enthusiastically adopted by myself on this side of the water. The disgusting deformity resulting from this treatment we all admitted, and for years it was a question, both with these patients and the profession, whether the operation would survive or the patients be left to struggle along without it till the disease carried them off. A patient who had submitted to it was a curiosity, and students came from long distances to see its performance. Surgeons asserted they would rather die than submit to it themselves, and refused to recom-

mend it.   Nevertheless, we had nothing better, and gradually it came to be universally
admitted that the patients were better after
its performance, that they gained in flesh, were
relieved of their pain, were often very comfortable in spite of the deformity, and the operation became a well-recognized method of
treatment, and has so remained.   In fact it has
constantly increased in popularity with the
profession, until now it has reached the opposite extreme, and these patients instead of being curiosities have become only too sadly
common.

I speak very advisedly, for too many have
come to me to be relieved of their cure, where
the condition certainly never justified such
radical treatment.   What was never more than
a choice between life and death, or between
life with a disgusting deformity and a life of
constant suffering, has become a routine practice in all cases of severe ulceration; and the
only reason why more patients have not been
operated upon is because, I am glad to say, it
is still very hard to gain their consent.   The

operation has been done without accurate diagnosis; it has been done below the disease instead of above in cases of ordinary intestinal catarrh; it has been done for constipation instead of stricture, and these patients are much worse off for their surgery. It has in fact met the fate which seems to attach to any operation, at the present time, which is comparatively simple of technique, and has once become recognized.

At the time when this operation was first advocated, in spite of all the objections to it, we had nothing better, as I have said. But, meanwhile, another method of treatment has been steadily forcing itself into notice, and it is one which, on account of the inherent difficulties of its performance, will never meet the same fate as has colostomy and suffer from too much popularity. I refer to the removal of the entire rectum, exactly as we would remove diseased ovaries or the uterus.

When Bryant advised colostomy in the treatment of these cases the operation of excision was still in its infancy and attended by a very high death-rate—nearly thirty-three per cent. It was only employed in cases of cancer, and the results were not very satisfactory even in that. But this also has grown in favor as the technique has improved and the mortality been lowered, until now, with experienced operators, the mortality is scarcely, if at all, greater than was that of the operation for making an artificial anus at that time. This has been accomplished by constant study of several men, all working in the same line toward simplicity and perfection of technique and result; and the operation, although one of the most difficult in surgery, is no longer any more dangerous when done by an experienced operator than many which are of daily occurrence in this room, in the department of diseases of women.

I say when done by experienced operators, and I mean to insist a little upon this, point. Most of you who have once seen the operation performed have little desire to attempt it, because its inherent difficulties are evident. My own death-rate when I began with it about ten years ago was the same as that of the German surgeons with whom the procedure originated. This has been gradually falling, until now I have lost but one in over two years, or in about twenty-five cases, and this is due to no essential change in the operation itself, but merely to greater facility in its performance, which has come from practice. The operation formerly took two or even three hours and was often fatal from shock and loss of blood; it now seldom takes more than half an hour, and the hemorrhage is seldom serious. Nurses who used to tell me they had never seen a case recover have grown accustomed to seeing these patients leave the hospital cured in a few weeks, and with no more apparent risk of serious results than in many laparotomies. In my last case the patient never had any shock, the pulse was as good at the end of the operation as at the commencement, there was never more than one degree rise of temperature, and the patient was up and convalescent at the end of a fortnight. This cannot be accomplished by any man, no matter how good a surgeon he may be, in his first series of cases, and can only come from practice. At first it will take you also two or three hours to complete the operation and your patients also will die of shock.

And now the question for us to decide is, by which of these two operations shall we treat this woman.

Were the disease with which we have to deal a cancer of the bowel, instead of what it is, other factors would come in which make the rule of practice easy to decide. If there is hope of cure we remove the disease. If too late for this we make an artificial anus. In the latter case we know that life cannot be prolonged many years at the best, and that the operation will give the greatest relief with the least possible risk. For such cases the operation will always be indicated. But this is not cancer, and this woman will not die if we make an artificial anus. On the contrary, she has as good a chance of living to old age as any of us, and the deformity we inflict will always remain. At first she may be grateful, but after a time, as her present sufferings disappear, she will forget her present condition, while the appreciation of just what we have done for her will grow. She will find herself doomed for life to an unnatural condition for a disease which would not probably ever have been fatal, and she may think we have not accomplished very much after all. She will not be cured, the old disease will still be there and still annoy her, she will not suffer as she did, but she may be an object of disgust to herself and her family, and she will come back and want to know if the disease itself cannot be cured and the artificial opening in the bowel closed again. This is no unusual circumstance; and really the abandonment of the disease to its course and the mere getting over its worst effects by opening the bowel above it is not so much of a surgical triumph as it may at first seem.

Moreover, the artificial anus is in itself a source of great annoyance in very many cases. The bowel protrudes and becomes eroded, hernia occurs through the incision, the discharge from the ulcerated rectum frequently escapes by this opening, and many of the patients are as anxious to be relieved of this condition as they were originally to be relieved of their disease.

On the other hand a successful extirpation with a good functional use of the parts after healing leaves the patient in a much better condition. She is cured of the disease by its complete removal; there is no longer any pain or discharge, and the parts are left as nearly as is possible as nature formed them. That there will be power to control the passages should never be promised, but this power will in the majority be greater than is possessed by those with an artificial anus in the left side, and the deformity will be infinitely less. The control of the passages, you must bear in mind, is generally lost in great measure before these patients come to the point of being operated upon, and they often tell me that there is more power in this regard after than before the operation.

This, then, is the operation I shall perform upon this woman rather than to make an arti-

ficial anus in the side, and this is the treatment,
I believe, which should now be the rule in all
these cases. It has taken me years to reach this
decision, which may seem a bold one to you, so
easy is one operation and so difficult the other;
but the fact that one is simple and the other
complicated has no bearing upon the question.
I am recommending not what is the easiest of
performance, but what gives much the best
local result. While the mortality of the one
I am going to advise was what it used to be
we were compelled to do the safer operation,
regardless of the deformity. With the reduc-
tion of the risk to what it now is we are, I believe, for the first time at liberty to recom-
mend the operation which gives the best result.

Again, this mortality, small as it now is, will
gradually become less, and the technique will
still further improve till the functional result
will be still better.

The operation of colostomy will not disappear, but it will not be taught as the proper
treatment of these cases and will be confined
to cases of cancer too advanced for complete

extirpation; and this, I think, will be much better for the public.

As to the technique of the operation it is unnecessary to give a description. I have described my own method very fully in my large work, with the necessary anatomical plates and every detail, and many of you have seen or will see its actual performance here. The two points to be acquired are speed in operating and the avoidance of hemorrhage, and these can only come to one who has acquired the facility which comes from actual practice. After these comes the proper suturing of the cut end of the bowel to the skin, which should always be at the site of the natural anus, and not, as it was at first, over the sacrum, for this deformity is but little better than an opening in the left groin.

The mortality is dependent absolutely upon the hemorrhage and shock, and upon the perfection of antisepsis, and, for this reason, the assistants should be men of experience in this particular operation, one of whom I always endeavor to have with me in every case. All sorts of unexpected complications are liable to arise, each of which you must be ready to meet on the instant, and no amount of reading will enable you to meet them as will the simple experience of having met the same thing before. When I say that I have seen good general surgeons and brilliant operators begin this operation and abandon it because of unexpected difficulties for which they were entirely unprepared, you will see why I do not advise you to undertake it.

To advocate such an operation as this in the place of one so simple as colostomy may seem to you advanced ground to take, but it is nevertheless the ground toward which I believe we have been gradually advancing ever since this operation became known. It is not so many years since we amputated above the knee in cases of disease of that joint, and the operation was a confession of our inability to cure. Now we excise the diseased joint itself. So here we have been practically amputating the bowel above the disease until the time has come to attack the disease itself.

Both the excision of the knee-joint and the excision of the rectum produce a certain amount of deformity, but not as great as in amputation, and in every way the result is much better. Now an amputation for a chronic inflammation of the knee-joint would be unjustifiable and the operation would only be indicated in cancerous disease. So I hope it will soon be in these cases.

I remember when it required a good deal of courage to do a colostomy, and when I dreaded the time when the patient would fully understand the condition in which he or she had been placed for life. For this reason I have always been slow in advising it, and have never performed it until life without it became intolerable, except in cases of cancer, where it distinctly retards the disease and should therefore be done as early as possible. But my own conservatism does not seem to have made much impression upon those who have seen the operation, and there is now a considerable demand from patients to be relieved of a condition for which I cannot find any justification.

I do not refer to cases such as the one before us, but to others where it would seem the operation must have been undertaken without sufficient thought or accurate diagnosis.